王大伟 著
雨青工作室 绘
[加]Jennifer May等 译

中国水利水电出版社
www.waterpub.com.cn
· 北京 ·

Nursery Rhyme

You say one, I count one the baby can wear pass-me-down clothes.
You say two, I count two, teach sharing with friends, so this he knows.
You say three, I count three, eating by myself is good.
You say four, I count four, learning to write as he should.
You say five, I count five, be brave and needles don't make me cry.
You say six, I count six, we help hurt animals, at least we try.
You say seven, I count seven, falling I can get up without help, I am strong.
You say eight, I count eight, sit on my chair and wash my socks, it doesn't take long.
You say nine, I count nine, before meals and after going potty always wash your hands, remind me.
You say ten, I count ten, my mother should teach me to recite poems, maybe three.

Good habits are more important than academic achievements.

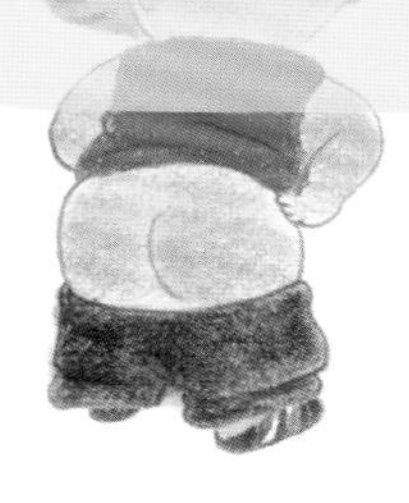
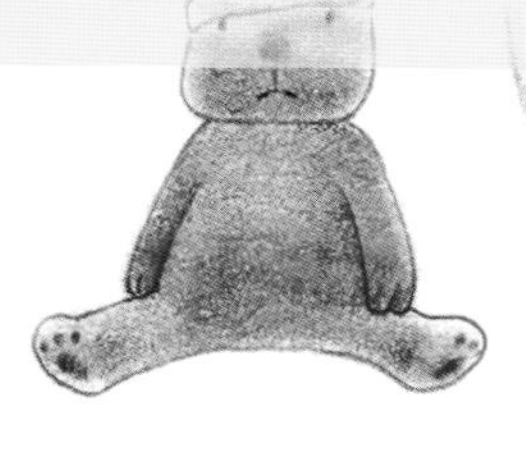

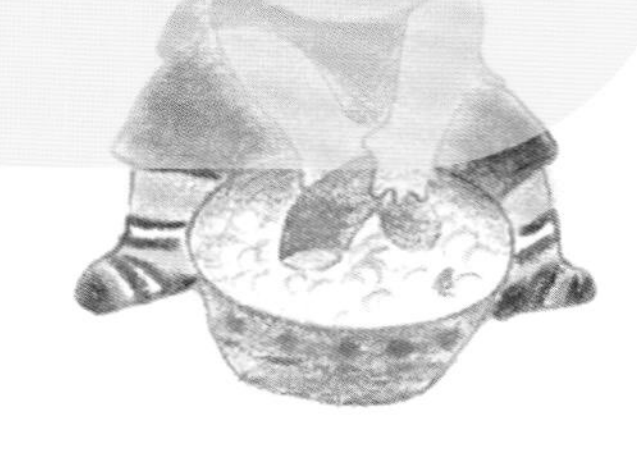

育儿新歌

你拍一，我拍一，宝宝要穿百家衣。

你拍二，我拍二，玩具让给小伙伴。

你拍三，我拍三，从小自己会吃饭。

你拍四，我拍四，一笔一画学写字。

你拍五，我拍五，打针不哭小老虎。

你拍六，我拍六，动物受伤咱们救。

你拍七，我拍七，摔倒爬起靠自己。

你拍八，我拍八，搬个板凳洗袜袜。

你拍九，我拍九，饭前便后要洗手。

你拍十，我拍十，妈妈教我背唐诗。

小案例
王大伟提示

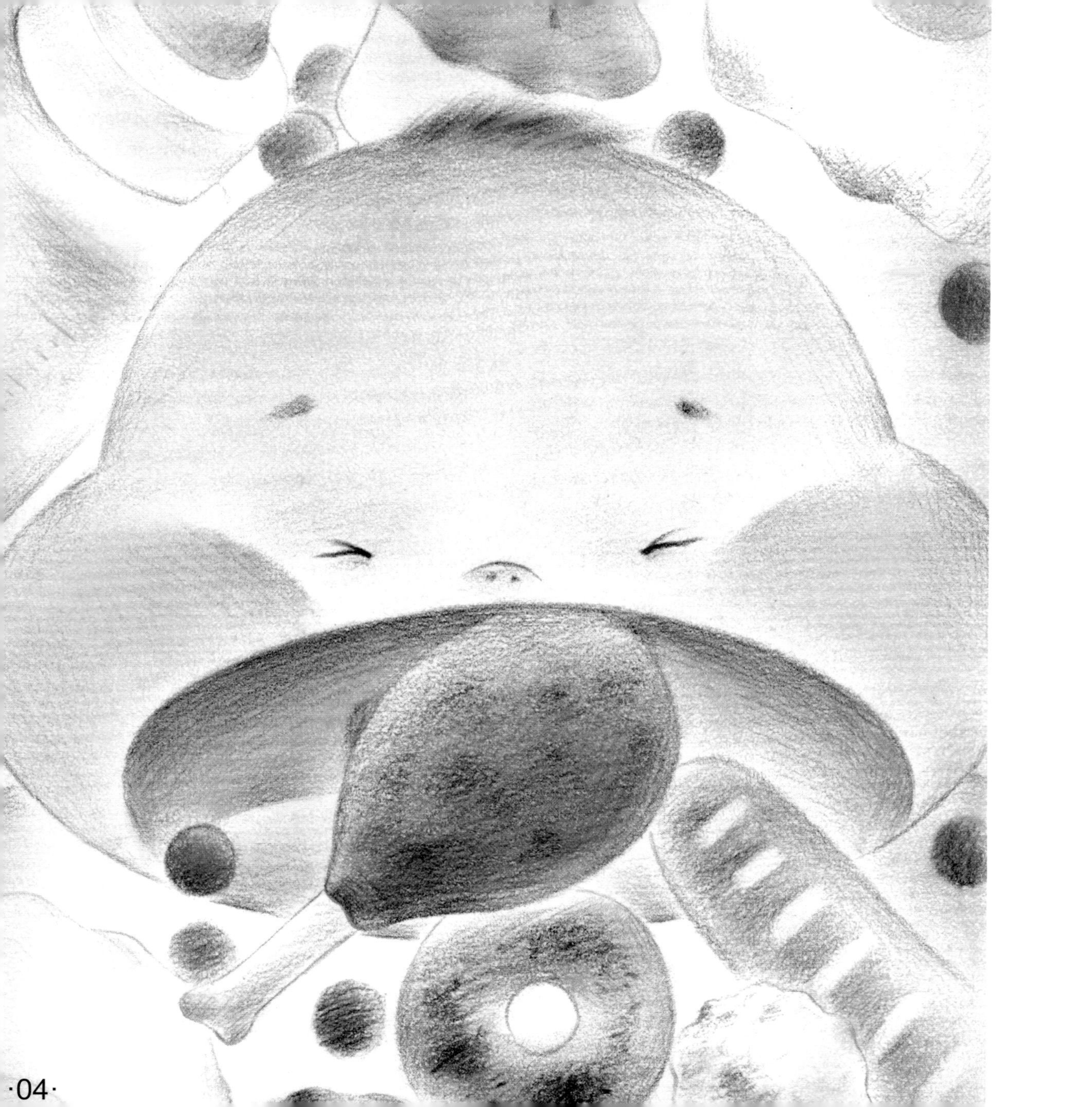

育婴秘法

光绪年《小儿推拿广义》中有一首乳子调护歌

养 子 须 调 护，看 承 莫 纵 驰。

乳 多 终 损 胃，食 壅 即 伤 脾。

被 浓 非 为 益，衣 单 正 所 宜。

无 风 频 见 日，寒 暑 顺 天 时。

Infant Feeding

Less milk, proper clothes, more daylight,
Follow and the day will be bright,
These are the three Chinese secrets;
If you follow, everything is alright.

There is an ancient book in China which records three parenting skills on infant feeding: "Drink less milk, wear proper clothes and see more daylight."

宝宝常遇仨坏蛋

宝宝常遇仨坏蛋，

感冒发烧扁桃腺。

打针吃药看医生，

父母不要太挂念。

小案例
王大伟提示

Babies Get Sick

Fever, tonsillitis, cold and my nose runs.
I can get those three bad ones.
See a doctor and take medicine;
Parents, it happens tons.

Cold, fever, and tonsillitis are common diseases. Parents don't need to be too anxious but should take children to the doctor in time.

妈妈也生病

妈妈怎么也生病?

床前坐个小板凳,

妈妈妈妈想吃啥?

宝宝送药沏杯茶。

小案例
王大伟提示

Mom is Sick

I sat near my mom's bed, so still.
Why does my mom get ill?
I can learn to care for her;
Give her a cup of tea and a pill.

Filial piety and caring are virtues.

节俭

爸爸挣钱不简单，

碗中米粒要吃干。

帮助贫困小朋友，

旧衣洗净也好穿。

小案例
王大伟提示

Being Thrifty

Any food leftover from the night before,
Eat it the next day and don't waste it anymore.
Wash my old clothes, be in a good mood,
And I will always help the poor.

There is an old saying in China: "Boys have to be raised in poverty". Thrifty is a virtue.

祖父教我写字

小猴子，爱模仿。

一幅字，挂墙上。

文房四宝都备好，

从小学字有榜样。

小案例
王大伟提示

Calligraphy

Learning the four treasures of study.
I love to imitate grandpa, my buddy.
Writing brush, inkstone, rice paper, and ink;
Good place, calm and clean, not cruddy.

Children should learn traditional calligraphy from an early age. The four treasures of the study are writing brush, inkstone, rice paper, and ink.

坏 叔 叔

鸟 妈 妈， 被 吓 走，

六 只 蛋， 窝 中 留。

来 了 一 个 坏 叔 叔，

个 个 打 碎 都 不 留。

小案例
王大伟提示

Bad Uncle

I have a bad uncle, you could say.
The mother bird flew away;
Six eggs in the nest smashed,
And anger all around that day.

Be kind,don’t learn that bad uncle who smashed all eggs in the nest.

坐不住的小猴子

小猴子，爱爬墙，

打碎碗，掉进缸。

带到医院化验锌，

调皮孩子查健康。

小案例
王大伟提示

Little Monkey

The little monkey,
climbed the wall, so spunky;
Fell into a jar, to the hospital;
For a zinc test, too much energy.

Although ADHD is rare, parents should pay attention to it.

小 虎 子

小 虎 子， 特 别 棒，

饿 上 一 顿 也 无 妨。

玩 具 要 比 邻 居 少，

常 穿 哥 哥 旧 衣 裳。

Little Tiger

I can save money, it’s a deal.
I can eat less at every meal.
Have fewer toys than friends;
Being thrifty, this is how I feel.

Children should be taught to inherit the tradition of diligence and thrift.

男孩要有爷们样

男孩要有爷们样，

吃苦节俭铁骨肠。

摔倒百次自爬起，

不让泪水挂脸庞。

小案例
王大伟提示

The Boy

I am a boy with a dream, not a bluff.
I will never give up, it is enough,
To be thrifty and exercise;
And get back up when things get tough.

Boys should learn to be a man.

小 公 主

小 公 主， 一 朵 花，

人 见 人 爱 骄 惯 她。

传 说 女 孩 要 富 养，

过 头 反 而 害 爸 妈。

A little Princess

I won’t be spoiled and feel sour.
I am a little princess, a lovely flower.
I don’t need money spent on me;
I know this will give me power.

Girls should not be spoiled.

天鹅飞上蓝天空

丑小鸭，安徒生，

长的丑，家里穷。

从小励志有爱心，

天鹅飞上蓝天空。

The Swan

I can become a white swan as a male.
Like the ugly duckling in the fairy-tale.
I love and don't listen to the thoughts of others.
I can become that swan with no fail.

The story of Andersen tells us that an ugly duckling can turn into a swan.

Safety Song

(1) Don't dress me, don't dress me.
Too hot or warm, too hot or warm
Not too much food this should be the norm.
Not too much food this should be the norm.
Don't dress me, don't dress me.
(2) I can play, I can play.
Eat by myself, eat by myself.
All these things I can do by myself.
All these things I can do by myself.
I can play, I can play.
(3) What I want, what I want.
I don't get it; I don't get it.
And we see all the good that it will bring.
And we see all the good that it will bring.
What I want, what I want.
(4) I can learn, I can learn.
All by myself, all by myself.
The best part is I learned something new.
The best part is I learned something new.
I can learn, I can learn.
(5) I can grow, I can grow.
Grow on my own, grow on my own.
Love me, punish me, so I am shown.
Love me, punish me, so I am shown.
I can grow, I can grow.

There is an old Chinese saying:
The boy can eat a thousand kinds of bitterness,
the girl can embroider ten thousand flowers.

小儿平安歌

若要小儿安，常带三分饥和寒。

若要小儿安，学会吃饭自己玩。

若要小儿安，少买玩具学勤俭。

若要小儿安，自己学习少充填。

若要小儿安，鼓励惩戒不能偏。

狗剩石头名字好，不求大贵求平凡。

男孩要吃千般苦，礼貌孝顺更为先。

父母爱多是暴力，学会放手难上难。

小案例
王大伟提示

在家篇

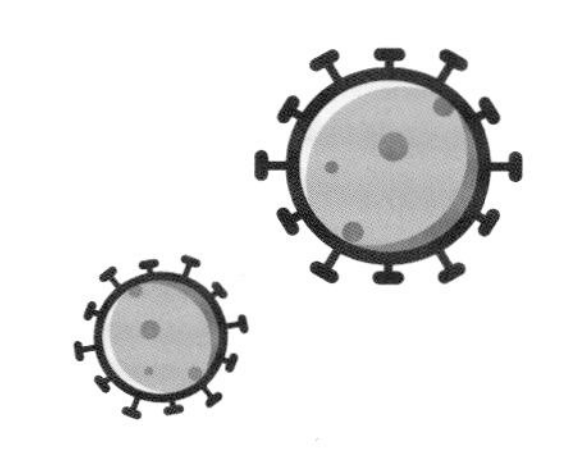

一、眼耳鼻口小病毒

眼睛的危险

1. 玩针线（预防扎伤眼睛）
2. 玩医院的注射器与针头
3. 玩具枪对着小朋友的眼睛
4. 剪刀不是平头的
5. 照相看闪光灯
6. 独自放鞭炮无大人看管
7. 固定的姿势看电视（预防斜眼）

耳朵的危险

1. 耳朵里放玩具的小零件
2. 放鞭炮时不捂耳朵、不远离
3. 铃铛与报警器放在耳边
4. 常挖耳屎不是好习惯
5. 洗澡时耳朵进水（要用棉签擦干）

鼻子的危险

1. 玩具有可拆解下的小零件（易误入鼻孔或误食）
2. 闻花（花籽易吸入气管）
3. 触摸到豆子、药片等（易误入鼻孔或误食）
4. 触摸到刺激性液体，酱油、醋放在孩子手边
5. 新装修的房子有异味，化学物质超标（预防白血病）
6. 雾霾天气开窗（可买空气净化器）

二、家中小病毒

1. 桌子上放水果刀
2. 家中有玻璃鱼缸（滑倒撞上很危险）
3. 家中挂有吊兰（下落伤人）
4. 楼房窗子没有安装高置插销（孩子可爬上去）

嘴巴的危险

1. 接触扣子与药片
2. 接触灯泡、体温计（易误食玻璃或水银）
3. 桌子上摆放花生和瓜子
4. 躺着吃东西
5. 吃果冻（易卡嗓子）
6. 笑谈、打闹时吃东西
7. 用饮料瓶子装洗涤剂、硫酸或毒药（易误饮误食）

三、水火小病毒

火的危险

1. 玩火柴
2. 点燃的蜡烛、蚊香靠近窗帘、蚊帐等可燃物品，不放在专用的架台上
3. 在床底，阁楼找东西时用油灯、蜡烛、打火机等明火照明
4. 阳台上、楼道内烧纸片或燃放烟花爆竹
5. 从燃放烟花爆竹区穿过
6. 厨房有自动打火煤气（大人不在家时，应关闭煤气）

水的危险

1. 随便玩水龙头
2. 盆浴先放热水（应先放凉水，防止孩子跳入）
3. 浴室没有防滑垫
4. 家附近有水坑、池塘或露天厕所
5. 雨季家附近有正在挖掘的建筑工地（积水易淹死人）

四、个人卫生小病毒

1. 饭前和便后不洗手
2. 打喷嚏时不用手帕或纸巾捂嘴
3. 瓜果没洗干净就吃
4. 饭有怪味还吃
5. 吃饱了还吃零食
6. 扁豆没炒熟（防止中毒）

图书在版编目（CIP）数据

王大伟儿童安全童谣. 教子秘诀 : 汉英对照 / 王大伟著. -- 北京 : 中国水利水电出版社, 2021.9
ISBN 978-7-5170-9617-7

Ⅰ. ①王… Ⅱ. ①王… Ⅲ. ①安全教育－儿童读物－汉、英 Ⅳ. ①X956-49

中国版本图书馆CIP数据核字(2021)第093351号

责任编辑 李格（1749558189@qq.com 010-68545865）

书　　名	王大伟儿童安全童谣：教子秘诀 WANG DAWEI ERTONG ANQUAN TONGYAO：JIAOZI MIJUE
作　　者	王大伟 著
绘　　图	雨青工作室
英文翻译	[加]Jennifer May 王大伟 陈诗楠 刘原
配音朗读	王许疃 李晟元 郑方允 崔瓔峤 侯清芸 吴郁暖 钟璇 郑淑予
出版发行	中国水利水电出版社 （北京市海淀区玉渊潭南路1号D座 100038） 网址：www.waterpub.com.cn E-mail：sales@mwr.gov.cn 电话：（010）68367658（营销中心）
经　　售	北京科水图书销售中心（零售） 电话：（010）88383994、63202643、68545874 全国各地新华书店和相关出版物销售网点
排　　版	韩雪
印　　刷	天津久佳雅创印刷有限公司
规　　格	210mm×190mm 24开本 5印张（总） 120千字（总）
版　　次	2021年9月第1版 2021年9月第1次印刷
总 定 价	68.00元（全4册）